Quaternary Quotient

Dividing Time by Ice Ages

Obadiah Carter

Chapter 1: The Quaternary Clock Setting the Stage

Deep Time and the Rise of the Quaternary

The vastness of geologic time is almost beyond human comprehension. Imagine Earth's 4.6-billion-year history compressed into a single year. On this scale, the dinosaurs, those seemingly ancient behemoths, only appear in late December, and their reign ends just before New Year's Eve. Humans? We wouldn't even register until the last few seconds of the year. This immense timescale, encompassing epochs upon epochs, is what geologists call Deep Time. It's within this vast expanse that we find the Quaternary Period, a mere blip on the geological clock, yet a period of immense significance, particularly for us.

The Quaternary, spanning the last 2.6 million years, might seem insignificant compared to the hundreds of millions of years occupied by the dinosaurs or the billions of years before their arrival. However, this period marks a dramatic shift in Earth's history, a time when our planet's climate became dominated by a cycle of glacial expansion and retreat. This is why the Quaternary is often referred to as "The Ice Age," though it encompasses a series of glacial cycles, not just one prolonged freeze.

The transition into the Quaternary wasn't abrupt. It was a gradual process driven by a complex interplay of geological and astronomical factors. Tectonic plates continued their slow dance, rearranging continents and altering ocean currents. These shifts, in turn, influenced global atmospheric and oceanic circulation patterns, setting the stage for dramatic climate fluctuations.

Simultaneously, subtle changes in Earth's orbit around the sun, known as Milankovitch Cycles, began to exert a more pronounced influence on our planet's climate. These cycles, occurring over tens of thousands of years, affect the amount and distribution of solar radiation reaching Earth's surface, acting as a pacemaker for the advance and retreat of massive ice sheets.

The onset of the Quaternary and its characteristic ice ages ushered in a period of dramatic environmental change, particularly across the higher latitudes. Vast ice sheets, sometimes miles thick, advanced and retreated across continents, carving out valleys, grinding down mountains, and leaving behind a mosaic of glacial landforms. Sea levels fell during glacial periods as water became locked away in massive ice sheets, exposing land bridges and connecting continents. Conversely, interglacial periods, like the one we live in today, saw the retreat of glaciers, rising sea levels, and the return of warmer temperatures.

These dramatic environmental fluctuations had profound implications for life on Earth. The Quaternary became a time of great evolutionary pressure, forcing plants and animals to adapt to constantly shifting conditions. Some species migrated vast distances to find suitable habitats, while others developed specialized adaptations to survive the cold.

This period witnessed the rise of iconic Ice Age megafauna like woolly mammoths, giant sloths, and saber-toothed cats, creatures specifically adapted to the challenges of a glacial world.

The Quaternary also marks a pivotal chapter in human evolution. It was during this time, amidst the challenges and opportunities presented by a fluctuating climate, that our own species, Homo sapiens, emerged. The need to adapt to changing environments likely spurred the development of key human traits like complex social structures, tool use, and the ability to exploit a wide range of resources.

Understanding the Quaternary is not merely an academic exercise in studying the distant past. It provides crucial context for understanding the present and future of our planet. The Quaternary serves as a stark reminder of the dynamic nature of Earth's climate and the profound influence it has on life. By studying the causes and consequences of past climate change, we gain valuable insights into the potential impacts of current and future warming trends. The Quaternary, with its cycles of ice and change, holds valuable lessons for our own time, a time when humanity faces the challenge of a rapidly changing climate.

Defining an Ice Age Glacial Cycles and Climate Shifts

The term "Ice Age" conjures images of a world locked in perpetual winter, a frozen wasteland where glaciers reign supreme. While evocative, this picture is not entirely accurate. An Ice Age, or more accurately, a glacial age, is a period in Earth's history when global temperatures are cold enough to allow for the formation of extensive ice sheets and glaciers, covering significant portions of the planet's surface. However, these periods are not one long, continuous freeze. Instead, they are characterized by a cyclical pattern of glacial expansion and retreat, a rhythmic pulse between colder glacial periods and warmer interglacial periods.

These cycles, often spanning tens of thousands of years, are driven by a complex interplay of factors, with variations in Earth's orbit around the sun playing a key role. These orbital variations, known as Milankovitch Cycles, influence the amount and distribution of solar radiation reaching Earth's surface, acting like a celestial thermostat. When the orbital configuration favors cooler summers and reduced solar radiation at high latitudes, snow and ice accumulate more readily during the winter months than they melt in the summer. Over time, this imbalance leads to the growth of massive ice sheets.

These growing ice sheets, in turn, trigger a cascade of feedback mechanisms that amplify the cooling trend. As ice expands, it reflects more sunlight back into space, further reducing the amount of solar radiation absorbed by the Earth's surface. This increased reflectivity, known as the albedo effect, contributes to a positive feedback loop, accelerating the cooling process and driving the planet deeper into a glacial period.

However, the Earth's climate system is dynamic and complex. While orbital forcing provides the initial trigger, other factors, such as changes in atmospheric greenhouse gas concentrations, ocean circulation patterns, and volcanic activity, also influence the duration and intensity of glacial cycles. These factors can either amplify or dampen the effects of orbital forcing, leading to variations in the length and severity of glacial and interglacial periods.

During a glacial period, the impact on Earth's environment is profound. Ice sheets, sometimes miles thick, expand across continents, carving out valleys, grinding down mountains, and leaving behind a legacy of glacial landforms. These massive ice sheets lock away vast amounts of water, causing sea levels to drop dramatically, sometimes by hundreds of feet. Coastlines are redrawn, land bridges emerge, connecting continents and allowing for the migration of plants and animals.

Conversely, interglacial periods, like the one we currently inhabit, are marked by the retreat of glaciers, rising sea levels, and warmer temperatures. These periods provide opportunities for life to flourish, with ecosystems adapting to the changing conditions. Forests expand, ice-covered regions become habitable, and biodiversity increases.

It's important to remember that even within glacial and interglacial periods, climate is not static. Shorter-term climate fluctuations, often lasting for centuries or even decades, are superimposed on the longer glacial-interglacial cycles. These fluctuations, driven by a variety of factors, including volcanic eruptions, solar variability, and changes in ocean currents, can cause significant regional and global temperature swings.

Understanding these glacial cycles and climate shifts is crucial for comprehending the dynamic nature of Earth's climate system. By studying the past, we gain valuable insights into the forces that have shaped our planet and continue to influence our present and future. The Quaternary, with its dramatic swings between glacial and interglacial conditions, serves as a powerful reminder of the Earth's climatic variability and the profound impact it has on life.

The Quaternary Toolkit Dating Methods and Proxies

Imagine trying to piece together the story of a long-lost civilization with only a handful of scattered artifacts and weathered ruins. That's the challenge facing Quaternary scientists as they attempt to reconstruct Earth's climate and environmental history over the past 2.6 million years. Fortunately, they have a toolkit at their disposal, a collection of ingenious methods and techniques for deciphering the clues hidden within the Earth's archives.

One essential tool in this endeavor is dating methods, techniques that allow scientists to establish a chronological framework for past events. These methods, like detectives dusting for fingerprints, help determine the age of geological formations, fossils, and other materials, providing a timeline for the Quaternary.

Radiocarbon dating, perhaps the most well-known dating method, relies on the decay of a radioactive isotope of carbon, carbon-14, to determine the age of organic materials up to around 50,000 years old. This method has revolutionized our understanding of recent Earth history, allowing scientists to date everything from ancient trees and charcoal deposits to the remains of extinct Ice Age animals and early human settlements.

For older materials, other dating methods, such as uranium-series dating and potassium-argon dating, are employed. These methods, based on the decay of radioactive isotopes with much longer half-lives, can date materials millions of years old, providing insights into the deep time history of the Quaternary.

However, dating methods alone only tell part of the story. To reconstruct past environments and climate, scientists rely on proxies, indirect indicators of past conditions preserved in the natural world. These proxies, like puzzle pieces scattered across time, provide clues about past temperatures, precipitation patterns, vegetation, and even atmospheric composition.

One of the most valuable archives of past climate information lies buried within ice sheets. Ice cores, extracted from glaciers and ice caps, contain trapped air bubbles, tiny time capsules preserving ancient air. By analyzing the composition of these bubbles, scientists can determine past atmospheric greenhouse gas concentrations, providing a direct window into the Earth's ancient atmosphere.

Ocean sediments, another rich source of proxy data, accumulate over millennia, layer upon layer, like pages in a history book. These sediments contain a wealth of information, including the fossilized remains of microscopic marine organisms called foraminifera. The chemical composition of these fossils, particularly the ratio of different oxygen isotopes, reflects the temperature and salinity of the ocean water in which they lived, providing insights into past ocean conditions and global climate.

On land, fossil pollen grains, preserved in lake sediments and peat bogs, offer a glimpse into past vegetation and climate. Different plant species have distinct pollen shapes, allowing scientists to identify the types of plants that once grew in a particular region. By analyzing changes in pollen assemblages over time, scientists can track the expansion and contraction of forests, grasslands, and other ecosystems in response to climate change.

Another intriguing proxy lies hidden within caves in the form of speleothems, mineral formations that grow slowly over time, drip by drip. These formations, including stalactites and stalagmites, contain layers of calcium carbonate, a mineral that incorporates isotopes of oxygen and carbon from the surrounding environment. By analyzing the isotopic composition of these layers, scientists can reconstruct past temperatures, rainfall patterns, and even vegetation changes.

Dating methods and proxies, combined, provide a powerful toolkit for reconstructing past environments and understanding Earth's climate history. By carefully analyzing the clues preserved in ice cores, ocean sediments, fossil pollen, speleothems, and other archives, scientists can piece together a detailed picture of the Quaternary, revealing the dynamic interplay of climate, geology, and life over the past 2.6 million years. This understanding of the past is not merely an academic exercise; it provides a crucial context for interpreting present-day environmental changes and anticipating future climate trends.

The Players on Stage Climate Geology and Life Intertwined

Earth's climate is not a solitary actor on a global stage. It's the product of a complex and dynamic interplay between a cast of interconnected players, each with its own role to play in shaping the planetary environment. Climate, geology, and life, far from being independent entities, are engaged in a constant dance, influencing and responding to each other over timescales ranging from millennia to moments.

Geology, the very foundation upon which life exists, provides the stage for this intricate performance. The arrangement of continents, driven by the slow but relentless movement of tectonic plates, dictates the flow of ocean currents, influencing heat transport across the globe. Mountain ranges, thrust upward by colliding plates, alter atmospheric circulation patterns, creating rain shadows and influencing regional climates. Volcanic eruptions, those fiery outbursts from Earth's interior, inject massive amounts of gases and particles into the atmosphere, sometimes causing short-term cooling but also contributing to the long-term greenhouse effect.

Climate, in turn, shapes the geological landscape. Glaciers, those rivers of ice, carve out valleys, grind down mountains, and deposit vast quantities of sediment, leaving an indelible mark on Earth's surface. Rainfall patterns dictate the flow of rivers, shaping canyons and floodplains, while weathering and erosion, driven by temperature and precipitation, slowly sculpt the land over millennia.

Life, the most dynamic player on this planetary stage, both responds to and influences climate and geology. Plants, through photosynthesis, draw down carbon dioxide from the atmosphere, influencing the greenhouse effect and playing a crucial role in regulating Earth's climate. The expansion and contraction of forests, driven by climate change, alter the reflectivity of Earth's surface, influencing regional and global temperatures. Marine organisms, from microscopic algae to coral reefs, play a vital role in the carbon cycle, influencing atmospheric carbon dioxide levels over geological timescales.

The Quaternary Period, with its dramatic swings between glacial and interglacial conditions, provides a stark example of this interconnectedness. Changes in Earth's orbit, those subtle shifts in our planet's celestial dance, triggered a cascade of effects that rippled through the climate system. As ice sheets expanded, they altered ocean currents, influenced atmospheric circulation patterns, and drove changes in vegetation, further amplifying the cooling trend. These changes, in turn, profoundly impacted the distribution and evolution of life, leading to the extinction of some species and the rise of others.

The story of the Quaternary is not merely one of climate change; it's a testament to the interconnectedness of Earth's systems. The rise and fall of ice sheets, the migration of plant and animal communities, the evolution of new species – all are intertwined with the dynamic interplay of climate, geology, and life.

This interconnectedness is not confined to the past; it's a defining feature of our planet's present and future. As human activities increasingly alter the Earth's climate system, we're witnessing the consequences of this interconnectedness firsthand. Melting glaciers, rising sea levels, shifting weather patterns – these are not isolated events but symptoms of a planet in flux, a planet where climate, geology, and life are responding in concert to the pressures of human-induced change.

Understanding this interconnectedness is not merely an academic exercise; it's crucial for navigating the challenges of the Anthropocene, this new epoch in Earth's history shaped by human activity. As we grapple with the consequences of a changing climate, we must remember that we're not just dealing with a single variable but with a complex web of interactions. Solutions to the climate crisis will require a holistic approach, one that recognizes the interconnectedness of Earth's systems and seeks to address the root causes of environmental change. By understanding the intricate dance between climate, geology, and life, we can begin to chart a more sustainable course for our planet and ourselves.

A Glimpse into the Quaternary World

Step back in time, just two and a half million years, and you find yourself in a world both familiar and strikingly different from our own. The continents, though holding roughly their present positions, are sculpted by vast ice sheets, their edges sprawling across the Northern Hemisphere. Mammoths and mastodons roam across frozen landscapes, while saber-toothed cats stalk their prey amidst grasslands stretching as far as the eye can see. This is the Quaternary Period, a time of dramatic climate fluctuations, marked by the ebb and flow of massive glaciers and the resilience of life adapting to ever-changing conditions.

The Quaternary, encompassing the Pleistocene and the current Holocene epochs, is characterized by a cyclical pattern of glacial expansion and retreat, driven by variations in Earth's orbit and amplified by feedback mechanisms within the climate system. During glacial periods, vast ice sheets, sometimes miles thick, blanketed much of North America, Europe, and Asia, locking away immense quantities of water and causing sea levels to plummet by hundreds of feet. Coastlines were redrawn, exposing land bridges that connected continents and allowed for the migration of plants and animals.

Imagine standing on the Bering Land Bridge, a vast expanse of tundra connecting Siberia and Alaska, a migratory highway for mammoths, bison, and even early humans. The air is biting, the landscape stark but beautiful, a testament to the power of nature to reshape the world.

These glacial periods, though harsh, were not devoid of life. In fact, they fostered the evolution of unique ecosystems and remarkable creatures adapted to the cold. Woolly mammoths, with their thick fur coats and massive tusks, are iconic symbols of these glacial landscapes, their remains unearthed from permafrost, frozen in time. These giants shared their icy realm with woolly rhinoceroses, giant sloths, and fearsome predators like saber-toothed cats and dire wolves, their adaptations a testament to the ingenuity of life in the face of environmental challenges.

But the Quaternary is not just a story of ice and cold. Interspersed between these glacial periods were warmer interglacials, times of glacial retreat, rising sea levels, and flourishing ecosystems. These periods, like the one we currently inhabit, provided opportunities for life to diversify and expand. Forests replaced ice sheets, grasslands gave way to woodlands, and biodiversity flourished.

During these interglacials, early humans thrived, their populations expanding as they migrated across continents, following the ebb and flow of glaciers and the movements of their prey. These were times of innovation and cultural development, as humans crafted new tools, developed sophisticated hunting strategies, and created enduring works of art on cave walls, capturing the essence of their world.

The Quaternary, with its dramatic climate swings and the resilience of life in the face of change, offers a unique perspective on our planet's dynamic history. It's a reminder that change is a constant in Earth's story, and that life, though challenged, has an extraordinary capacity to adapt and thrive. By studying this period, we gain valuable insights into the forces that have shaped our planet and continue to influence our present and future. The fossils of extinct creatures, the remnants of ancient ice sheets, the geological formations carved by glaciers – these are not just relics of the past; they're pieces of a puzzle that help us understand the interconnectedness of Earth's systems and the profound impact of climate change on life.

Chapter 2: Frozen Earth Unveiling the Glacial Cycles

Milankovitch Cycles The Orbital Engine of Ice Ages

Earth's climate, far from being static, dances to the rhythm of subtle yet powerful cycles, a celestial waltz orchestrated by our planet's orbital variations. These variations, known as Milankovitch cycles, named after the Serbian geophysicist who first described them, are the metronome of Earth's glacial-interglacial heartbeat, influencing the timing and intensity of ice ages over millennia.

Imagine Earth's orbit, not as a perfect circle, but as a slightly elongated ellipse, its shape shifting subtly over time. This variation in eccentricity, as it's known, influences the amount of solar radiation Earth receives at different points in its orbit. When the orbit is more elliptical, the difference in solar energy received between Earth's closest and farthest points from the sun is greater, potentially influencing the onset or demise of ice ages.

But Earth's orbital dance doesn't end there. Our planet also wobbles on its axis, like a spinning top slowing down, a phenomenon known as axial precession. This wobble, with a cycle of approximately 26,000 years, alters the timing of the seasons relative to Earth's position in its orbit. Currently, Earth's axis points towards Polaris, the North Star, but over millennia, it will point towards other stars, changing the timing of summer and winter in each hemisphere and influencing the distribution of solar energy across the globe.

Adding further complexity to this celestial ballet is the tilt of Earth's axis, known as obliquity. This tilt, currently at about 23.5 degrees, is what gives us our seasons. However, this angle is not constant; it varies between approximately 22.1 and 24.5 degrees over a cycle of roughly 41,000 years. A greater tilt means more extreme seasons, with hotter summers and colder winters, while a smaller tilt results in more moderate seasonal differences.

These three orbital parameters – eccentricity, precession, and obliquity – interact in a complex dance, their combined effects influencing the distribution and intensity of solar radiation reaching Earth's surface. These subtle variations in solar energy, though seemingly small, can have a profound impact on Earth's climate system, particularly over long periods.

During periods when orbital configurations favor cooler summers and milder winters in the Northern Hemisphere, snow and ice that accumulate during the winter months are less likely to melt entirely during the summer. This leads to a gradual buildup of ice sheets over time, reflecting more sunlight back into space and further amplifying the cooling trend. This positive feedback loop, where a cooler climate leads to more ice, which in turn leads to further cooling, is a key driver of glacial expansion.

Milankovitch cycles provide the framework for understanding the timing of glacial-interglacial cycles over the Quaternary Period. Geological evidence, from ice cores to ocean sediments, reveals a striking correlation between variations in Earth's orbit and the ebb and flow of ice sheets. These cycles, however, are not the sole drivers of climate change. Other factors, such as volcanic eruptions, variations in solar output, and, more recently, human activities, can also influence Earth's climate system, sometimes amplifying or mitigating the effects of orbital forcing.

Understanding Milankovitch cycles is not just an academic exercise; it's crucial for understanding Earth's climate past, present, and future. These cycles tell us that Earth's climate is naturally dynamic, subject to variations over long timescales. While they operate over tens of thousands of years, far too slow to explain current warming trends, they provide a crucial context for understanding the sensitivity of Earth's climate system to even subtle changes in energy balance.

As we grapple with the challenges of human-induced climate change, understanding the role of Milankovitch cycles and other natural climate forcings is essential for disentangling the complex interplay of factors shaping our planet's climate. By studying the past, we gain valuable insights into the forces that have shaped our planet and continue to influence its future.

Feedback Loops Amplifying the Glacial Signal

Earth's climate system is a delicate dance of interconnected parts, where a change in one component can trigger a cascade of effects, amplifying or dampening the initial disturbance. This intricate web of interactions is characterized by feedback loops, cyclical processes that can either amplify or diminish the effects of climate forcings, such as those triggered by Milankovitch cycles.

Imagine a snowball rolling down a hill, gathering more snow and momentum as it goes. This is analogous to a positive feedback loop, where an initial change triggers a series of events that reinforce the initial change, leading to a runaway effect. In the case of Earth's climate, positive feedback loops play a crucial role in amplifying the relatively small changes in solar radiation caused by Milankovitch cycles, pushing the planet into and out of glacial periods.

One of the most powerful positive feedback loops in the climate system involves ice albedo, the reflectivity of Earth's surface. Ice, being bright white, reflects a significant portion of incoming solar radiation back into space. As ice sheets expand during glacial periods, Earth's overall albedo increases, reflecting more sunlight away and further cooling the planet. This cooling, in turn, leads to further ice sheet growth, amplifying the initial cooling effect and driving further glacial expansion.

But ice albedo is not the only player in this glacial drama. Another crucial feedback loop involves the concentration of carbon dioxide (CO_2) in the atmosphere. During glacial periods, as temperatures drop and ice sheets expand, the amount of CO_2 in the atmosphere decreases. This decrease in CO_2, a potent greenhouse gas, further weakens the greenhouse effect, amplifying the cooling trend initiated by orbital forcing.

The reasons for this decrease in atmospheric CO2 during glacial periods are complex and not fully understood, but they likely involve changes in ocean circulation patterns, biological productivity, and the balance between carbon uptake and release by terrestrial ecosystems. Regardless of the precise mechanisms, the net effect is a reduction in atmospheric CO2, further amplifying the cooling trend.

These positive feedback loops, involving ice albedo and atmospheric CO2, work in concert to amplify the relatively small changes in solar radiation caused by Milankovitch cycles. They act like a volume knob on Earth's climate system, turning up the cooling effect during glacial periods and amplifying the warming trend during interglacials.

But Earth's climate system is not solely governed by positive feedback loops. Negative feedback loops, which act to dampen or stabilize the climate system, also play a crucial role. For example, as temperatures rise, the atmosphere can hold more moisture, leading to increased cloud cover. Clouds, depending on their type and altitude, can either reflect sunlight back into space, cooling the planet, or trap heat, warming it. This cloud feedback loop can act as either a positive or negative feedback, depending on the prevailing conditions, adding complexity to the climate system's response to forcings.

Understanding feedback loops is crucial for comprehending the sensitivity of Earth's climate system and its response to both natural and human-induced changes. While Milankovitch cycles provide the initial trigger for glacial-interglacial cycles, it's the interplay of positive and negative feedback loops that amplifies or dampens these initial forcings, shaping the magnitude and duration of these climatic swings. As we continue to alter Earth's atmosphere through the burning of fossil fuels and other activities, it's more important than ever to understand how these feedback loops will respond, shaping the future of our planet's climate.

Ice Sheet Dynamics Growth Flow and Retreat

Ice sheets, those colossal rivers of ice that carve mountains and reshape continents, might seem like static giants, frozen in time. But beneath their seemingly immobile surfaces lies a dynamic world of flow and change, driven by a delicate balance between growth, movement, and retreat.

Imagine a giant pancake batter slowly spreading out on a griddle, its edges advancing outward as more batter is added to the center. This is akin to how ice sheets grow and flow. Snowfall accumulates in the central, higher-elevation regions of an ice sheet, compacting under its own weight to form ice. As the ice thickens, it begins to flow outward under the force of gravity, spreading laterally across the landscape.

This flow is not uniform. Ice sheets, like rivers, flow faster in the center and slower at the edges, where friction with the underlying terrain increases. This differential flow creates crevasses, deep fissures in the ice that can extend for miles, a testament to the immense forces at play. The speed of ice flow also varies depending on the slope of the terrain, the temperature of the ice, and the presence of meltwater at the base of the ice sheet, which can act as a lubricant.

As ice sheets flow outward, they encounter various obstacles, from mountain ranges to valleys, which can influence their shape and movement. Ice can surge over mountains, creating spectacular icefalls, or it can be channeled through valleys, carving deep fjords into the landscape. These interactions between ice and topography create the dramatic landscapes we associate with glaciated regions.

The growth of an ice sheet depends on the balance between snowfall, which adds mass to the ice sheet, and ablation, the loss of ice through melting, sublimation (the direct transition of ice to water vapor), and calving, the breaking off of icebergs at the edges of the ice sheet. When snowfall exceeds ablation, the ice sheet grows; when ablation exceeds snowfall, the ice sheet shrinks.

During glacial periods, when temperatures are colder and snowfall increases, ice sheets expand, their edges advancing across vast distances. These advancing ice sheets act like giant bulldozers, scraping away vegetation, carving out valleys, and transporting huge amounts of rock and sediment, which are deposited at the ice sheet's margins as moraines, marking the extent of their former reach.

As temperatures warm during interglacial periods, the balance shifts, and ablation begins to exceed snowfall. Ice sheets retreat, their edges melting back, exposing the land that was once buried beneath thousands of feet of ice. This retreat is not always a gradual process; ice sheets can experience periods of rapid collapse, driven by positive feedback loops, such as the ice albedo feedback, where melting ice exposes darker surfaces, which absorb more solar radiation, leading to further melting.

The dynamics of ice sheet growth, flow, and retreat are complex, influenced by a multitude of factors, including climate, topography, and the internal properties of the ice itself. Understanding these dynamics is crucial for predicting how ice sheets will respond to future climate change and the potential consequences for sea level rise. As we continue to warm the planet through the burning of fossil fuels, we are altering the delicate balance that governs ice sheet dynamics, with potentially profound implications for our planet and its inhabitants.

Glacial Landforms A Legacy Etched in Stone

Across the Earth's surface, like wrinkles on an ancient face, lie the indelible marks of past glaciations. These glacial landforms, etched in stone and sediment, are silent but powerful testaments to the immense power of ice to sculpt and reshape the landscape. They stand as natural archives, preserving within their forms a record of past climates and glacial processes, waiting to be deciphered by those who know how to read their language.

Imagine a giant rasp, scraping across the land, carving out valleys and leaving behind a trail of debris. This is the legacy of valley glaciers, rivers of ice that flow down mountainsides, carving out U-shaped valleys, their characteristic signature. These valleys, with their steep, often sheer walls and broad, flat bottoms, contrast sharply with the V-shaped valleys carved by rivers, a testament to the erosive power of moving ice.

As glaciers advance, they pluck rocks from the valley walls, incorporating them into their icy grip. These rocks, dragged along the glacier's base, act like chisels, carving striations, or scratches, into the bedrock, providing clues to the direction of ice flow. The size and shape of these striations can tell us about the size and velocity of the glacier that created them.

At the glacier's snout, where the ice melts as fast as it advances, it deposits its load of rock and sediment, forming a terminal moraine, a ridge of debris that marks the glacier's furthest advance. These moraines, often composed of a jumbled mix of boulders, gravel, and sand, can be several hundred feet high and stretch for miles, forming natural dams that can create lakes.

As glaciers retreat, they leave behind a landscape littered with other telltale features. Erratics, boulders of rock that differ in composition from the underlying bedrock, stand as lonely sentinels, transported miles from their source by the ice. Drumlins, streamlined hills sculpted by the ice, resemble giant teardrops, their long axes pointing in the direction of ice flow.

Kettles, depressions formed by the melting of blocks of ice buried in glacial debris, dot the landscape, often filled with water to form kettle lakes. Eskers, sinuous ridges of sand and gravel, mark the former courses of subglacial rivers, which flowed beneath the ice, carrying away meltwater and sediment.

Beyond these large-scale features, glacial landscapes are often characterized by a subtle but pervasive micro-relief. Glacial polish, a smooth, almost mirror-like surface, results from the abrasion of bedrock by fine-grained sediment embedded in the ice. Roche moutonnée, asymmetrical bedrock knobs, sculpted by the passage of ice, resemble sheep lying down, their gentle slopes facing up-glacier and their steeper, plucked sides facing down-glacier.

These glacial landforms, from towering mountains to subtle bedrock engravings, are not merely scenic wonders; they are invaluable tools for reconstructing past climates and understanding the dynamics of ice sheets. By mapping the distribution of glacial features, scientists can determine the extent of past glaciations, the direction of ice flow, and the timing of glacial advances and retreats.

Moreover, the study of glacial landforms provides insights into the processes that shape Earth's surface, from erosion and deposition to the interplay between climate, topography, and ice dynamics. These landforms, sculpted by the relentless power of ice, stand as a testament to the dynamism of our planet and the profound influence of climate on the landscape we see today.

Chapter 3: Landscapes in Flux Shaping the World

Glacial Erosion Carving Mountains and Valleys

Imagine a world where mountains crumble, valleys deepen, and landscapes transform under the relentless advance of ice. This is the world of glacial erosion, where the seemingly static power of ice acts as a sculptor of continents, carving mountains and valleys over millennia.

Glaciers, despite their slow and steady movement, are incredibly powerful agents of erosion. They possess a unique ability to carve and shape the Earth's surface, not just through their immense weight but also through the processes of plucking, abrasion, and meltwater erosion.

Picture a glacier flowing over a landscape, its massive tongue of ice pressing down with unimaginable force. As it moves, it freezes to the bedrock below, grasping onto fractured rock and sediment. This process, known as plucking, is akin to a giant hand picking up loose stones. As the glacier continues its slow march, it tears these rocks from the Earth, carrying them along and leaving behind a scarred and fractured landscape.

But glacial erosion is not just about brute force. As the glacier drags these plucked rocks across the bedrock, they act like sandpaper on a cosmic scale, grinding and smoothing the surface below. This process, called abrasion, leaves behind telltale signs of the glacier's passage: glacial striations. These scratches, etched into the bedrock, run parallel to the direction of ice flow, providing clues to the glacier's path and power.

The erosive power of glaciers isn't limited to the ice itself. Meltwater, produced by the pressure of the overlying ice and by warmer temperatures at the glacier's base, plays a significant role in shaping glacial landscapes. This meltwater, often flowing under high pressure, carves channels and tunnels within and beneath the glacier, further accelerating erosion.

These subglacial rivers, laden with sediment, act like high-pressure jets, carving deep gorges and valleys into the bedrock. When these subglacial channels are eventually exposed by the glacier's retreat, they leave behind dramatic canyons and valleys, often characterized by steep walls and flat bottoms, a testament to the erosive power of meltwater.

The combined forces of plucking, abrasion, and meltwater erosion are responsible for some of the most awe-inspiring landscapes on Earth. The majestic U-shaped valleys, with their steep walls and wide, flat bottoms, are a hallmark of glacial erosion, contrasting sharply with the V-shaped valleys carved by rivers.

Fjords, those deep, narrow inlets of the sea flanked by towering cliffs, are another stunning example of glacial handiwork. These dramatic features, found in high-latitude regions, were carved by glaciers that extended out to sea, their immense weight carving deep troughs into the Earth's crust.

From the rugged peaks of the Alps to the deep fjords of Norway, glacial erosion has left an enduring mark on our planet. These landscapes, shaped by the slow but relentless power of ice, are not just aesthetically breathtaking; they serve as a constant reminder of the dynamic forces that have shaped, and continue to shape, the Earth's surface. As we marvel at the beauty of a glacier-carved valley or the serene waters of a fjord, we are witnessing the legacy of a powerful force, a force that has sculpted mountains, carved valleys, and left an indelible mark on our planet's history.

Glacial Deposition Building Plains and Moraines

Imagine a sculptor, not chiseling away at stone, but rather, meticulously piecing together a landscape from a jumble of materials. This is the essence of glacial deposition, a process where glaciers, having carved and sculpted the Earth's surface, lay down their burden of rock, sediment, and debris, building plains, hills, and ridges that reshape the face of the planet.

Glaciers, for all their erosive power, are also great hoarders. As they advance across the landscape, they gather up a motley assortment of materials—from fine silt and clay to massive boulders—all carried along in the glacier's icy grip. This collection of debris, known as glacial till, is a testament to the glacier's journey, a tangible record of the rocks and sediment it has scraped, plucked, and transported along the way.

As glaciers melt and retreat, they leave behind this accumulated debris, like a messy child abandoning a pile of toys. This process of glacial deposition creates a variety of landforms, each with its own unique characteristics and story to tell.

Perhaps the most prominent features of glacial deposition are moraines, those sinuous ridges of till that mark the boundaries of past glaciers. Like lines drawn on a map, moraines delineate the extent of former ice sheets and glaciers, providing clues to their size, shape, and movement.

Terminal moraines, marking the furthest advance of a glacier, stand like ramparts, often forming natural dams that trap meltwater, creating lakes and wetlands. Lateral moraines, formed along the edges of valley glaciers, snake along the valley sides, resembling long, earthen embankments.

But glacial deposition isn't limited to these large-scale features. Outwash plains, vast expanses of sediment deposited by meltwater flowing away from glaciers, spread out like aprons in front of melting ice sheets. These plains, often characterized by braided rivers and fertile soils, are testaments to the power of meltwater to transport and deposit vast quantities of sediment.

Within these outwash plains, one often finds kettles, depressions formed by the melting of blocks of ice buried in the glacial debris. These kettles, ranging in size from small ponds to large lakes, add a touch of whimsy to the landscape, their still waters reflecting the sky above.

Drumlins, those enigmatic, teardrop-shaped hills, are another intriguing product of glacial deposition. These streamlined hills, often found in clusters, are thought to be formed by the reworking of glacial till beneath flowing ice, their smooth, elongated shapes aligned with the direction of ice flow.

Eskers, sinuous ridges of sand and gravel, snake across the landscape, marking the former courses of subglacial rivers. These ridges, often elevated above the surrounding terrain, provide valuable clues to the drainage patterns beneath past glaciers.

Glacial deposition, then, is not merely a process of dumping debris; it's a creative force, shaping and reshaping the Earth's surface in a myriad of ways. These landforms, built from the remnants of eroded mountains and valleys, are not static monuments to the past; they are dynamic features, constantly evolving under the influence of wind, water, and vegetation.

The legacy of glacial deposition is all around us, from the fertile soils of the Midwest to the rolling hills of New England. These landscapes, sculpted by the dual forces of erosion and deposition, are a testament to the enduring power of ice to shape our world, leaving behind a tapestry of landforms that continue to inspire and intrigue us today.

Periglacial Processes The Freeze Thaw Cycles Impact

Beyond the icy grip of glaciers, a subtle but powerful force shapes the landscape: the relentless cycle of freezing and thawing. This periglacial environment, a zone where temperatures hover around the freezing point of water, is a realm of constant change, where the expansion and contraction of ice wreaks havoc on the ground, creating a unique and often dramatic landscape.

Imagine water seeping into cracks in the soil or bedrock. As temperatures plummet, the water freezes, expanding with an irresistible force. This expansion, repeated countless times over centuries, widens cracks, shatters rocks, and heaves the ground into a chaotic jumble of broken fragments.

This freeze-thaw cycle, a hallmark of periglacial environments, is the driving force behind a suite of processes that shape the landscape in subtle and dramatic ways. One of the most visible manifestations of this process is frost heaving, where the repeated freezing and thawing of the ground pushes stones and soil upwards, creating patterned ground.

These patterned grounds, often taking the form of polygons, circles, or stripes, are a testament to the power of ice segregation. As water freezes in the soil, it draws water from surrounding areas, creating lenses of ice that grow larger with each freeze-thaw cycle. These ice lenses push the soil upwards, creating the distinctive patterns that characterize periglacial landscapes.

Another striking feature of periglacial environments is solifluction, a slow, downslope movement of waterlogged soil. Imagine a hillside, saturated with water from melting snow or permafrost. As temperatures fluctuate around freezing, the ground thaws and freezes repeatedly, causing the soil to lose its structural integrity and flow like a viscous fluid.

This solifluction, often imperceptible to the naked eye, can transport vast amounts of soil and sediment over time, creating distinctive lobes, terraces, and other flow features on the landscape. These features, often draped across hillsides like a rumpled blanket, are a testament to the slow but persistent power of gravity and the freeze-thaw cycle.

Periglacial processes are not limited to the surface; they also play a crucial role in shaping the subsurface. The formation of permafrost, permanently frozen ground, is a defining characteristic of periglacial environments. This permafrost, often extending hundreds of feet below the surface, acts as a barrier to drainage, creating waterlogged soils and influencing vegetation patterns.

The thawing of permafrost, driven by climate change, can have dramatic consequences. As the ground thaws, it releases methane, a potent greenhouse gas, into the atmosphere, further accelerating climate change. The thawing of permafrost also weakens the ground, leading to landslides, subsidence, and damage to infrastructure.

From the patterned grounds of the Arctic tundra to the solifluction lobes of alpine meadows, periglacial processes shape a diverse and dynamic landscape. These processes, driven by the relentless cycle of freezing and thawing, are a testament to the power of even subtle temperature changes to transform the Earth's surface. Understanding these processes is crucial, not only for comprehending the evolution of

periglacial landscapes but also for predicting the
impacts of climate change on these sensitive
environments.

Sea Level Changes Drowning Coastlines and Creating Land Bridges

Imagine a world where coastlines shift dramatically,
where once-thriving coastal settlements vanish
beneath the waves, and land bridges emerge,
connecting continents and reshaping the distribution
of life on Earth. This is the world shaped by sea level
changes, a powerful force that has sculpted our
planet's coastlines for millennia.

Sea level, far from being constant, is a restless giant,
rising and falling over time in response to a complex
interplay of geological and climatic factors. These
fluctuations, occurring over time scales ranging from
decades to millions of years, have profound impacts
on coastal environments, submerging landmasses,
carving new shorelines, and influencing the course of
human history.

One of the primary drivers of sea level change is the
growth and decay of ice sheets. During glacial periods,
vast quantities of water are locked away in massive ice
sheets, causing sea level to plummet. These periods of
glacial advance, like the last glacial maximum some
20,000 years ago, saw sea levels drop by hundreds of
feet, exposing vast areas of continental shelves and
transforming the geography of our planet.

Conversely, during interglacial periods, like the one we are currently experiencing, rising temperatures cause ice sheets to melt, returning water to the oceans and causing sea levels to rise. This rise in sea level, though gradual, can have dramatic consequences for coastal communities, leading to increased erosion, flooding, and saltwater intrusion into freshwater aquifers.

But sea level changes are not solely driven by the waxing and waning of ice sheets. Tectonic activity, the restless movement of Earth's crustal plates, also plays a crucial role in shaping coastlines. The collision of tectonic plates can uplift coastal regions, exposing new land and pushing back the sea. Conversely, the subsidence of coastal regions, often associated with the movement of tectonic plates or the compaction of sediments, can lead to submergence and the loss of coastal land.

Volcanic eruptions, those fiery displays of Earth's internal heat, can also influence sea level, albeit on a more localized scale. Large eruptions, by spewing massive amounts of ash and debris into the atmosphere, can temporarily cool the planet, leading to increased ice formation and a slight drop in sea level.

The impacts of sea level changes are far-reaching, influencing not only coastal landscapes but also the distribution of plants and animals. During periods of low sea level, land bridges often emerge, connecting continents that were once separated by vast stretches of water. These land bridges, such as the Bering Land Bridge that once connected Asia and North America, allowed for the migration of plants and animals between continents, influencing the biodiversity of entire regions.

The rise and fall of sea level is etched into the very fabric of our planet's history, recorded in the sedimentary layers, ancient shorelines, and drowned landscapes that bear witness to these dramatic changes. Understanding the causes and consequences of sea level change is crucial, not only for comprehending the evolution of our planet but also for predicting and mitigating the impacts of future sea level rise on coastal communities and ecosystems.

The Quaternary Record Reading Landscapes

Imagine a species facing a changing world. Temperatures rise, rainfall patterns shift, and familiar landscapes transform. Survival hinges on finding new homes, safe havens where life can persist amidst the upheaval. These havens, known as refugia, play a crucial role in the story of life on Earth, offering sanctuary during times of environmental change and serving as springboards for the eventual recolonization of landscapes once the crisis has passed.

Refugia are pockets of relative stability, where climate and environmental conditions remain suitable for a species even as the surrounding environment undergoes dramatic shifts. These havens can take many forms: a high-altitude mountain valley shielded from rising temperatures, a north-facing slope retaining moisture in an increasingly arid landscape, or an isolated island spared from the ravages of a continental disease outbreak.

The importance of refugia lies not only in their ability to provide immediate shelter but also in their role as reservoirs of biodiversity. Within these pockets of stability, species can persist, preserving their genetic diversity and evolutionary potential for a time when conditions become favorable for expansion. Without refugia, many species would face extinction in the face of rapid environmental change.

Identifying and understanding refugia is crucial for conservation efforts, particularly in the face of accelerating climate change. By mapping these havens of biodiversity, we can prioritize their protection, ensuring that these vital reservoirs of life persist as sources of resilience for future generations.

But the story of life in a changing world is not just about seeking refuge; it's also about movement, about the epic journeys undertaken by species in search of suitable habitats. This process of migration, often spanning generations and traversing vast distances, is a testament to the resilience and adaptability of life on Earth.

Imagine a herd of caribou, their hooves drumming across the tundra, following ancient migratory paths in search of food and breeding grounds. Or picture a flock of songbirds, navigating by the stars, their journey spanning continents as they track the shifting seasons. These migrations, honed over millennia, are finely tuned to the rhythms of the natural world.

Yet, as climate change disrupts these rhythms, altering the timing of seasons, the availability of resources, and the suitability of habitats, many species are facing unprecedented challenges. Migratory pathways that once led to abundance are now fraught with obstacles, forcing species to adapt, to forge new routes, or to face the consequences of a changing world.

Understanding the patterns and drivers of migration is essential for conservation efforts. By mapping migratory routes, identifying critical stopover points, and mitigating threats along these pathways, we can help ensure the survival of migratory species in a rapidly changing world.

The story of refugia and migration is a story of resilience, of the remarkable ability of life to adapt and persist in the face of environmental change. By studying these processes, we gain a deeper appreciation for the interconnectedness of life on Earth and the importance of safeguarding the biodiversity that sustains us all. As we face the challenges of a changing world, the lessons learned from these ancient journeys will be more valuable than ever.

Chapter 4: Life on the Edge Adapting to a Changing World

Refugia and Migration Tracking Suitable Habitats

Imagine a vast, arid landscape stretching towards the horizon, punctuated by isolated pockets of greenery. These oases, clinging to life amidst the harsh surroundings, are refugia – havens for plants and animals struggling to survive in a changing world. As climate change intensifies, these refugia become increasingly vital, offering not just sanctuary, but also a glimpse into the resilience of life itself.

Refugia are pockets of relative stability in a changing environment. Picture a mountain range, its peaks trapping moisture and creating microclimates where temperatures remain cool despite a warming trend in the lowlands. Or envision a deep canyon, its shaded depths offering respite from the scorching sun. These pockets of favorable conditions become arks of biodiversity, sheltering species struggling to adapt to the rapid pace of change elsewhere.

But refugia are not static museums. They are dynamic stages where the drama of survival unfolds in real-time. Species pushed to their limits by shifting temperatures or dwindling water sources find refuge in these havens, their populations intermingling and evolving in response to the new challenges. The genetic diversity fostered within these refugia becomes a crucial resource, a reservoir of adaptability that could prove vital for the survival of species in a rapidly changing world.

Understanding the location and characteristics of refugia is crucial for conservation efforts. By identifying these havens, we can prioritize their protection, ensuring that they continue to provide sanctuary for vulnerable species. Imagine a team of scientists, armed with climate models and ecological data, meticulously mapping these refugia, their work laying the groundwork for targeted conservation strategies that safeguard these vital areas.

However, seeking refuge is only one part of the story. For many species, survival hinges on movement – on undertaking epic journeys in search of more suitable habitats as the climate changes. This is the phenomenon of migration, a testament to the adaptability and resilience of life on Earth.

Picture a herd of wildebeest, their hooves thundering across the savanna as they follow ancient migratory paths in search of fresh grazing lands. Or envision a cloud of monarch butterflies, their delicate wings carrying them thousands of miles across continents, their journey timed to coincide with the blooming of nectar-rich flowers. These migrations, honed over millennia, are intricate dances with the environment, finely tuned to the rhythms of the natural world.

Yet, as climate change disrupts these rhythms, throwing the timing of seasons out of sync and altering the availability of resources, many species find their migratory pathways disrupted. Imagine those same wildebeest, their migration hampered by a drought that has turned their once reliable watering holes into dust bowls. Or picture the monarch butterflies, their journey jeopardized by a late frost that has decimated the wildflowers they rely on for sustenance.

The challenges posed by climate change to migratory species are significant. But just as refugia offer hope for survival, so too does our understanding of migration patterns. By mapping these journeys, identifying critical stopover points, and working to mitigate the threats posed by habitat loss and fragmentation, we can help ensure that these ancient rhythms continue.

The stories of refugia and migration are ultimately stories of hope. They remind us that even in the face of unprecedented environmental change, life finds a way. By understanding and protecting these havens and pathways, we can help ensure that the tapestry of life on Earth continues to thrive, even as the planet transforms around us.

Evolution in Overdrive Speciation and Extinction Events

Evolution is often portrayed as a slow, gradual process, a stately dance of adaptation unfolding over eons. But the Quaternary Period, with its dramatic climate swings and rapidly changing environments, reveals a different face of evolution – a face of accelerated change, of rapid diversification, and of extinctions that reshaped the very fabric of life on Earth. This is evolution in overdrive, a high-stakes game of survival where the rules are constantly changing, and the stakes have never been higher.

Imagine a landscape transformed, not over millions of years, but over mere millennia. Glaciers advance and retreat, carving new paths for rivers and isolating populations. Sea levels rise and fall, redrawing coastlines and creating islands where once there were none. These rapid environmental shifts act as powerful selective pressures, favoring those organisms with the traits needed to survive and reproduce in the new conditions.

The result is a burst of evolutionary innovation. Species diversify, adapting to exploit new ecological niches. New traits emerge, honed by natural selection to provide an edge in the struggle for survival. The fossil record of the Quaternary is replete with examples of this accelerated evolution, from the emergence of cold-adapted mammoths and woolly rhinoceroses to the diversification of finches on the Galapagos Islands, their beaks evolving in response to changes in food availability.

But the Quaternary is also a story of loss, a time marked by extinction events that reshaped the planet's biodiversity. These extinctions, often coinciding with periods of rapid climate change, highlight the vulnerability of life to environmental upheaval, even as they underscore the resilience of those species that manage to survive.

The disappearance of megafauna, those iconic giants that once roamed the planet, is perhaps the most striking example of Quaternary extinctions. Picture vast herds of mammoths and mastodons, their footsteps shaking the ground, their long tusks clashing as they compete for mates. Or imagine the ground sloth, its massive claws raking leaves from towering trees, its slow, deliberate movements a stark contrast to the swift predators that stalk the undergrowth.

These magnificent creatures, along with saber-toothed cats, giant sloths, and dire wolves, vanished from the Earth during the late Pleistocene and early Holocene epochs, their extinction coinciding with a period of rapid warming and the spread of humans across the globe. While the exact causes of these extinctions remain a topic of debate, the interplay of climate change, human impacts, and ecological shifts likely played a significant role.

These extinctions serve as a stark reminder of the fragility of ecosystems and the interconnectedness of life on Earth. The loss of even a single species can have cascading effects throughout an ecosystem, disrupting food webs, altering vegetation patterns, and ultimately reshaping the landscape itself.

Yet, amidst these extinctions, there are also stories of resilience, of species that weathered the storms of change and emerged, transformed but triumphant. The survival of these lineages provides a testament to the enduring power of life to adapt and persist, even in the face of extraordinary challenges.

Understanding the interplay of speciation and extinction events during the Quaternary is crucial for understanding the patterns of biodiversity we see today. By studying the past, we gain insights into the factors that drive evolutionary change, the vulnerability of species to environmental upheaval, and the remarkable resilience of life on Earth. These lessons are more relevant than ever as we grapple with the challenges of a rapidly changing planet, a planet where the forces of evolution are once again in overdrive.

Megafauna Giants of the Ice Age World

The air crackles with a dry, frigid bite. Snow crunches underfoot, the silence broken only by the mournful howl of the wind across a vast, icy expanse. This is the Pleistocene epoch, the time of the last Ice Age, and the Earth belongs to giants. Towering mammoths carve paths through snowy forests, their colossal tusks gleaming under a pale sun. Shaggy-coated rhinoceros graze on windswept plains, their massive horns held high. The ground trembles beneath the weight of giant sloths, their claws leaving deep furrows in the frozen earth. These are the megafauna, the magnificent behemoths that ruled the Ice Age world.

Imagine a creature that dwarfed today's elephants, its body cloaked in thick fur, its curved tusks longer than a human is tall. This was the woolly mammoth, perhaps the most iconic of all Ice Age giants. These majestic herbivores roamed the steppe-tundra, their massive bodies adapted to withstand the harsh conditions. Their enormous tusks, used for defense, attracting mates, and even clearing snow to reach buried vegetation, were a testament to the power of natural selection in a challenging environment.

But the mammoths were not alone. Across the frozen landscapes of the Northern Hemisphere, another giant roamed: the woolly rhinoceros. Covered in thick, reddish-brown fur, this formidable creature possessed two massive horns, the front one a formidable weapon and tool, reaching over three feet in length. These solitary herbivores were well-equipped to survive the harsh winters, using their horns to sweep snow aside and uncover buried vegetation.

In warmer regions, a different kind of giant held sway. The giant ground sloth, standing nearly 20 feet tall when rearing up on its hind legs, was a sight to behold. These massive herbivores, with their long claws and powerful builds, could reach high into trees to strip branches of leaves, their size deterring even the largest predators. Their slow, deliberate movements and seemingly gentle nature belie their status as one of the most imposing creatures of their time.

These megafauna were not merely impressive spectacles; they were keystone species, playing vital roles in shaping the ecosystems they inhabited. Their grazing habits influenced plant communities, their movements created pathways through dense vegetation, and their dung provided nutrients that enriched the soil. The echoes of their presence reverberated throughout the food web, influencing everything from predator populations to the distribution of plant life.

Yet, by the late Pleistocene and early Holocene epochs, these giants began to disappear. The reasons for their extinction are complex and still debated, but a combination of factors likely played a role. Climate change, with its dramatic shifts in temperature and vegetation patterns, undoubtedly posed challenges. The arrival of humans, with their sophisticated hunting techniques and growing populations, added another layer of pressure. The loss of these megafauna marked the end of an era, a profound shift in the Earth's ecosystems from which they have yet to fully recover.

The study of Ice Age megafauna provides a window into a lost world, a world shaped by different environmental forces and inhabited by creatures of awe-inspiring size and power. Their extinction serves as a stark reminder of the interconnectedness of life on Earth and the profound impact that environmental change and human activities can have on even the most dominant of species. As we grapple with the challenges of a changing planet, the lessons learned from these giants of the past take on a new urgency, urging us to protect the biodiversity that remains and to tread carefully upon the Earth.

The Rise of Humanity A Quaternary Success Story

The savanna stretches out under a blazing sun, a tapestry of golden grasses and acacia trees. A group of figures moves across this landscape, their silhouettes stark against the setting sun. These are not the lumbering forms of megafauna that have long defined this land. They are smaller, more agile, their steps lighter yet purposeful. They carry with them not brute strength, but tools and intelligence, the hallmarks of a new kind of dominance. This is Homo sapiens, and their story during the Quaternary Period is one of remarkable expansion, adaptation, and ultimately, unprecedented impact on the planet.

The story of humanity's rise begins long before the Quaternary, but it is during this period, marked by fluctuating climate and changing environments, that our species truly flourishes. While other hominid species struggle to adapt, Homo sapiens demonstrate a remarkable capacity for innovation and resilience, their journey a testament to the power of adaptability in the face of environmental change.

Imagine early humans, huddled around flickering fires, their faces illuminated by the dancing flames. They share not just food and warmth, but also stories, knowledge passed down through generations, a testament to the growing power of language and social learning. These early humans are not just surviving; they are thriving, their populations expanding, their tools becoming more sophisticated, their mastery of fire opening up new possibilities for cooking, warmth, and protection.

As the Quaternary unfolds, humans begin to migrate out of Africa, their journeys taking them across continents, their footsteps eventually marking their presence on every corner of the globe. They adapt to new environments, from the frozen landscapes of the Arctic to the dense rainforests of the tropics, their ingenuity and adaptability allowing them to thrive in conditions that would have defeated other hominid species.

The development of increasingly sophisticated tools and hunting techniques allows humans to exploit a wider range of resources. They develop spears, harpoons, and eventually, bows and arrows, their hunting prowess increasing, their impact on prey species becoming more pronounced. They learn to utilize fire not just for warmth and cooking, but also as a tool for managing landscapes, clearing vegetation to create favorable hunting grounds and encourage the growth of desirable plants.

The rise of agriculture during the Holocene epoch marks a turning point in human history and in humanity's relationship with the natural world. No longer solely reliant on hunting and gathering, humans begin to actively shape their environment, clearing forests, cultivating crops, and domesticating animals. This newfound ability to control food production leads to settled lifestyles, larger populations, and the emergence of villages, towns, and eventually, cities.

The Quaternary Period, for all its environmental fluctuations and challenges, provides the crucible in which Homo sapiens is forged. It is a period that witnesses the blossoming of human ingenuity, the development of complex social structures, the birth of art and symbolism, and the dawn of civilization. Yet, this success story is not without its complexities. As humans rise to dominance, their impact on the planet becomes increasingly profound, ushering in a new era, the Anthropocene, where human activities become the dominant force shaping the Earth's ecosystems.

The story of humanity during the Quaternary is a testament to our species' adaptability, resilience, and capacity for innovation. It is a story that continues to unfold, a story that now carries with it the responsibility to use our intelligence and ingenuity not just for our own survival, but for the preservation of the planet and the myriad life forms with which we share it.

Chapter 5: Echoes of the Past Reconstructing Quaternary Environments

Ice Cores Frozen Archives of Climate Change

Deep within the polar regions, locked beneath layers of accumulated snow and ice, lie frozen libraries of Earth's history. These are not libraries of parchment and ink, but of ancient precipitation, compressed and crystallized over millennia – ice cores, silent sentinels that hold within their icy grip a detailed record of past climate change.

Imagine a core, extracted from the heart of an ice sheet, its length a timeline stretching back thousands of years. Each layer, like the rings of a tree, represents a period of snowfall, compressed and preserved over time. But these layers hold more than just frozen water. Trapped within them are tiny bubbles, each a miniature time capsule containing samples of the ancient atmosphere. As scientists drill deeper, extracting cores that extend further back in time, they are able to reconstruct a detailed history of Earth's climate, revealing the ebb and flow of greenhouse gases, the pulse of volcanic eruptions, and the subtle shifts in temperature that have shaped our planet.

The analysis of ice cores is a meticulous process, requiring specialized equipment and expertise. Scientists carefully extract gas bubbles from the ice, measuring their composition to determine the concentration of greenhouse gases, such as carbon dioxide and methane, present in the atmosphere at the time the ice was formed. These measurements provide a direct record of past greenhouse gas levels, allowing scientists to track their fluctuations over time and understand how they have influenced Earth's climate.

But the secrets held within ice cores go beyond gas bubbles. Dust particles, trapped within the ice, provide clues about volcanic eruptions, periods of drought, and even the intensity of solar activity. Volcanic eruptions, for example, release large amounts of ash and aerosols into the atmosphere, which can be transported by winds and deposited in polar regions, leaving a distinct fingerprint within the ice layers. Similarly, periods of drought can lead to increased dust transport, leaving a record of arid conditions in the ice.

By analyzing the isotopic composition of the ice itself, scientists can reconstruct past temperature variations. Water molecules containing different isotopes of oxygen, for example, have slightly different physical properties, and their relative abundance in ice layers reflects the temperature at the time the precipitation fell. By piecing together these isotopic records, scientists can create detailed temperature reconstructions, revealing the warmth of past interglacial periods and the chill of glacial advances.

Ice cores have provided some of the most compelling evidence for the link between greenhouse gases and climate change. Data from ice cores spanning hundreds of thousands of years clearly show a strong correlation between atmospheric carbon dioxide concentrations and global temperatures. The rapid increase in carbon dioxide levels since the Industrial Revolution, unprecedented in the past 800,000 years, stands out as a stark anomaly, a testament to the profound impact of human activities on Earth's climate system.

These frozen archives of climate change provide a unique perspective on the Earth's climate history, revealing the natural fluctuations that have shaped our planet over millennia. But they also serve as a stark warning, highlighting the unprecedented scale and pace of human-induced climate change. By studying the past, as recorded in the icy layers of polar ice caps, we gain a deeper understanding of the delicate balance of Earth's climate system and the urgent need to address the challenges of a warming world.

Ocean Sediments Unlocking Deep Time Secrets

The ocean floor, a realm of perpetual twilight, holds within its depths a treasure trove of Earth's history. Beneath the waves, far beyond the reach of sunlight, lie vast accumulations of sediment, layers upon layers of microscopic fossils, minerals, and other debris that have settled out of the water column over millions of years. These ocean sediments, like the pages of a vast and ancient book, record the ebb and flow of life, the vagaries of climate, and the dramatic geological events that have shaped our planet.

Imagine a research vessel, its lights cutting through the inky blackness of the deep ocean. On deck, scientists gather around a sediment core, freshly extracted from the ocean floor. This core, a cylindrical cross-section of the ocean floor, represents millions of years of Earth's history, each layer a time capsule containing clues about past environments and the organisms that inhabited them.

The analysis of ocean sediments is a multidisciplinary endeavor, drawing upon the expertise of geologists, paleontologists, chemists, and oceanographers. By studying the physical and chemical properties of the sediment, scientists can reconstruct past ocean conditions, such as temperature, salinity, and nutrient levels. The presence of certain minerals, for example, can indicate periods of intense volcanic activity, while variations in the ratio of different isotopes of oxygen can reveal past changes in ocean temperature and global ice volume.

But perhaps the most fascinating aspect of ocean sediments lies in the microscopic fossils they contain. The skeletal remains of foraminifera, diatoms, and other tiny marine organisms, preserved within the sediment, provide a window into past ecosystems and the evolution of life in the oceans. By identifying and dating these fossils, scientists can track the rise and fall of different species, the impact of environmental change on marine ecosystems, and the long-term patterns of evolution that have shaped the diversity of life on Earth.

One of the most remarkable discoveries to emerge from the study of ocean sediments is the evidence for catastrophic events that have punctuated Earth's history. The Chicxulub impact, for example, which wiped out the dinosaurs 66 million years ago, left a distinct signature in ocean sediments worldwide. A thin layer of iridium, an element rare on Earth but abundant in asteroids, marks the boundary between the Cretaceous and Paleogene periods, a testament to the devastating impact that ended the reign of the dinosaurs.

But ocean sediments also reveal a more gradual, yet equally profound, process that has shaped our planet: plate tectonics. By studying the magnetic properties of ocean sediments, scientists can reconstruct the history of Earth's magnetic field, which is generated by the movement of molten iron in the planet's core. These studies have provided compelling evidence for the theory of seafloor spreading, the process by which new oceanic crust is created at mid-ocean ridges and spreads outward, pushing continents apart.

The study of ocean sediments provides a unique and invaluable perspective on Earth's history, revealing the interplay of geological, biological, and chemical processes that have shaped our planet over millions of years. These deep-sea archives offer insights into the evolution of life, the dynamics of Earth's climate system, and the catastrophic events that have punctuated our planet's past, providing a rich tapestry of knowledge that helps us understand the Earth's past, present, and future.

Fossil Pollen Reconstructing Ancient Vegetation

Picture a landscape from a time long past, a tapestry of forests, meadows, and wetlands lost to the relentless march of time. How can we possibly reconstruct these vanished ecosystems, knowing that the plants themselves have long since decayed into dust? The answer lies in a remarkable, almost indestructible organic material: pollen. These tiny grains, produced by the male reproductive organs of flowering plants, hold the key to unlocking the secrets of ancient vegetation and understanding how ecosystems have changed over millennia.

Fossil pollen grains, preserved in sediments deposited in lakes, bogs, and even ocean floors, are surprisingly durable. Their outer walls, composed of a substance called sporopollenin, are resistant to decay, able to withstand the ravages of time far better than the plants that produced them. As a result, pollen grains can persist for thousands, even millions, of years, providing a remarkably detailed record of past vegetation.

Imagine a scientist peering through a microscope at a sample of sediment extracted from a lake bed. Each slide tells a story, revealing a miniature world of fossilized pollen grains. The scientist meticulously identifies each grain, its unique shape, size, and ornamentation hinting at its origin. Some grains, with their intricate patterns and delicate structures, might belong to trees that once towered over the landscape, while others, smaller and more simply shaped, might represent grasses and herbs that carpeted the forest floor.

By analyzing the abundance and diversity of pollen grains preserved at different depths within a sediment core, scientists can track changes in vegetation over time. A shift from forest to grassland pollen, for example, might indicate a period of drying climate, while an increase in the abundance of pollen from aquatic plants could suggest a rise in lake levels.

But fossil pollen can reveal much more than just changes in plant communities. It can also provide insights into past climates, human activities, and even the evolution of plant life itself. For example, by studying the distribution of pollen from different species, scientists can reconstruct past migration patterns of plants in response to climate change. The presence of pollen from cultivated plants, such as wheat or corn, can provide evidence of early agriculture and the impact of humans on the landscape.

One of the most valuable aspects of fossil pollen analysis is its ability to provide a long-term perspective on ecological change. By comparing modern pollen assemblages with those preserved in ancient sediments, scientists can track the long-term dynamics of ecosystems and understand how they have responded to past environmental changes. This information is crucial for predicting how ecosystems might respond to future climate change and for developing effective conservation strategies.

The study of fossil pollen is a testament to the power of observation and the interconnectedness of life on

Earth. From tiny grains, barely visible to the naked eye, scientists can reconstruct vanished landscapes, track the ebb and flow of plant life, and gain a deeper understanding of the complex interactions between climate, vegetation, and human activities that have shaped our planet over millennia. These tiny time capsules offer a window into the past, providing valuable insights that can help us navigate the environmental challenges of the future.

Speleothems Drip by Drip Unraveling the Past

The darkness inside a cave, often perceived as empty and still, is actually alive with unseen processes and whispers of bygone eras. Within these subterranean realms, nature patiently crafts intricate structures, drop by patient drop – speleothems. These formations, from the majestic stalactites that descend from the cave ceilings to the stalagmites that rise from the floor, are far more than just geological curiosities. They are, in fact, natural archives, silently recording the pulse of climate and environmental change over thousands of years.

Imagine venturing deep into a cave, the air cool and damp, the only light emanating from your headlamp. As you navigate the uneven terrain, you encounter a towering flowstone, its surface rippled like frozen water. This seemingly inert formation is a testament to the power of time and the persistence of natural processes. Each layer, deposited over centuries by mineral-rich water seeping through the cave ceiling, holds a unique chemical signature, a snapshot of the climate and environmental conditions prevalent during its formation.

The key to unlocking the secrets held within speleothems lies in their chemical composition. As water percolates through the soil and rock above a cave, it dissolves minerals, particularly calcium carbonate, carrying them in solution. When this water reaches the cave atmosphere, some of the dissolved carbon dioxide escapes, causing the calcium carbonate to precipitate out of solution, forming the intricate structures we know as speleothems.

The beauty of speleothems as climate archives lies in their ability to preserve a continuous, high-resolution record of environmental change. As each layer forms, it incorporates trace elements and isotopes present in the water, creating a chemical fingerprint of the prevailing conditions at the time. By analyzing the composition of these layers, scientists can reconstruct past variations in temperature, rainfall, and vegetation cover.

One of the most widely used techniques for dating speleothems is uranium-thorium dating. This method exploits the radioactive decay of uranium, which occurs at a known rate, into thorium. By measuring the ratio of these isotopes in a speleothem sample, scientists can determine its age with remarkable precision, often to within a few decades.

The information gleaned from speleothems has revolutionized our understanding of past climate change. For example, speleothem records from around the world have provided compelling evidence for abrupt climate shifts, such as the Younger Dryas event, a period of rapid cooling that occurred around 12,800 years ago. These records have also shed light on the relationship between climate change and human evolution, revealing how past climate fluctuations may have influenced the migration patterns and cultural development of our ancestors.

But speleothems are more than just passive recorders of climate change. They can also provide insights into other environmental processes, such as earthquake frequency and intensity. The impact of an earthquake can create fractures in the bedrock above a cave, altering the flow paths of water and leaving a distinct signature in the growth patterns of speleothems.

The study of speleothems is a testament to the ingenuity of scientists and their ability to extract valuable information from seemingly unlikely sources. These silent sentinels of the underworld, formed drop by patient drop over millennia, offer a unique and invaluable window into Earth's past, providing insights into the dynamics of climate change, the evolution of life, and the geological forces that have shaped our planet.

Dating the Past Establishing Chronologies

Time, an unceasing river, carries us forward, leaving behind a trail of moments both grand and subtle. Yet, the human desire to understand the past, to piece together the tapestry of history, necessitates a framework, a way to measure and order the events that have shaped our world. This is the essence of chronology, the science of dating the past, of establishing temporal relationships between events and placing them within the grand narrative of time.

Imagine an archaeologist carefully excavating a site, each layer of soil revealing fragments of pottery, tools, and other remnants of past lives. How can we determine the age of these artifacts, their place in the chronology of human history? This is where the ingenuity of scientific dating methods comes into play, offering a range of techniques that allow us to peer into the past and assign ages to events and objects.

One of the most widely recognized dating methods is radiocarbon dating, a technique that harnesses the natural decay of radioactive carbon-14. This isotope, formed in the upper atmosphere, is incorporated into all living organisms. When an organism dies, it ceases to take up carbon-14, and the existing amount within its remains begins to decay at a known rate. By measuring the remaining carbon-14 in a sample, scientists can estimate the time elapsed since the organism's death, providing ages for organic materials up to around 50,000 years old.

But what about events that predate the reach of radiocarbon dating, events that stretch back millions or even billions of years? For these ancient events, geologists turn to other radiometric dating methods, such as uranium-lead dating and potassium-argon dating. These techniques exploit the decay of radioactive isotopes with much longer half-lives, allowing scientists to date rocks and minerals that formed early in Earth's history.

Beyond radiometric dating, a suite of other methods provides complementary insights into the past. Dendrochronology, the study of tree rings, offers a remarkably precise dating tool for archaeological sites and environmental studies. Each year, trees lay down a new ring of growth, the width and composition of which reflect the prevailing climate conditions. By comparing tree-ring patterns from living and dead trees, scientists can create continuous chronologies that extend back thousands of years.

Another powerful dating tool, particularly useful for reconstructing past climates, is ice core dating. As snow accumulates in polar regions and high mountain glaciers, it traps air bubbles that preserve a record of past atmospheric composition. By analyzing the layers of ice, scientists can track changes in greenhouse gas concentrations, volcanic eruptions, and other climate-related events over hundreds of thousands of years.

The establishment of accurate chronologies is fundamental to our understanding of the past. It allows us to piece together the sequence of events that have shaped our planet and its inhabitants, from the formation of the continents to the rise and fall of civilizations. By placing events within a temporal framework, we can begin to unravel the complex interplay of geological, biological, and cultural processes that have driven change over time.

Dating the past is not merely an exercise in satisfying our curiosity about bygone eras. It also provides crucial context for understanding the present and anticipating the future. By studying past climate change, for example, we gain insights into the potential impacts of human activities on the Earth's climate system. Similarly, archaeological chronologies shed light on the long-term interactions between humans and their environment, offering lessons for sustainable living in the face of global challenges.

Chapter 6: The Anthropocene Humanitys Impact on the Quaternary

A New Epoch Defining the Human Fingerprint

The Earth, throughout its vast history, has been a silent witness to epochs of dramatic transformation. Continents have shifted, climates have fluctuated, and life has evolved from single-celled organisms to the rich tapestry of biodiversity we see today. Yet, never before has a single species wielded such influence over the planet, leaving an imprint so profound as to warrant the designation of a new epoch: the Anthropocene.

Imagine standing at the edge of a vast open-pit mine, a scar carved into the Earth's surface, a testament to the sheer scale of human activity. Across the globe, from the sprawling urban landscapes to the agricultural heartlands, our species has become a geological force, reshaping the planet at an unprecedented rate.

The concept of the Anthropocene, while still debated within scientific circles, encapsulates the profound impact of human activities on Earth's systems. It recognizes that our species has become a dominant force driving changes in the atmosphere, oceans, biosphere, and even the planet's geology.

One of the most compelling arguments for the Anthropocene lies in the dramatic alteration of

Earth's atmosphere. Since the Industrial Revolution, the burning of fossil fuels, deforestation, and other human activities have released massive amounts of greenhouse gases into the atmosphere, trapping heat and driving global warming at an unprecedented rate. The resulting climate change, with its rising sea levels, extreme weather events, and shifts in plant and animal life, is a stark reminder of our species' growing influence.

Beyond the atmosphere, the oceans, too, bear the imprint of human activity. Plastic pollution, a scourge of the modern era, has infiltrated even the deepest ocean trenches, posing a serious threat to marine life. Overfishing has decimated fish populations, disrupting marine ecosystems and jeopardizing a vital food source for billions of people.

The biosphere, the thin layer of life that envelops our planet, has also been profoundly altered by human actions. Habitat loss, fragmentation, and degradation, driven by agriculture, urbanization, and resource extraction, have pushed countless species to the brink of extinction, triggering what scientists believe is the sixth mass extinction event in Earth's history.

The very geology of the planet, once considered the realm of slow, inexorable processes, now bears the unmistakable mark of human activity. The sheer volume of concrete produced by our species, along with the proliferation of plastics, metals, and other synthetic materials, has created a distinct layer in the geological record, a testament to our industrial prowess and a lasting legacy of our presence on Earth.

The concept of the Anthropocene, while sobering, is not intended to evoke despair, but rather to foster a sense of responsibility and urgency. It is a call to action, a recognition that we have become the stewards of our planet, tasked with charting a more sustainable course for ourselves and future generations.

As we grapple with the challenges of the Anthropocene, it is essential to remember that we are not passive observers of these changes, but active participants in shaping the future of our planet. By embracing sustainable practices, reducing our environmental footprint, and fostering a deeper understanding of our interconnectedness with the natural world, we can strive to create a future where human ingenuity and environmental stewardship go hand in hand, ensuring a thriving planet for generations to come.

Climate Change Acceleration From Ice Ages to Global Warming

Earth's climate, far from being static, is a dynamic system, constantly responding to a complex interplay of natural forces. Over millennia, our planet has experienced dramatic swings in temperature, from frigid ice ages that entombed continents in ice to warmer interglacial periods like the one we inhabit today. Understanding these past climate shifts is crucial for contextualizing the unprecedented warming trend we are currently witnessing and for predicting the potential consequences of our actions.

Imagine a time when massive ice sheets, thousands of feet thick, stretched across vast swaths of the Northern Hemisphere, their icy fingers reaching as far south as present-day New York City. This was the scene during the Last Glacial Maximum, around 20,000 years ago, when Earth's average temperature was about 4-7 degrees Celsius cooler than today.

These cyclical ice ages, punctuated by warmer interglacial periods, are driven by variations in Earth's orbit and rotation, known as Milankovitch cycles. These subtle shifts in Earth's position relative to the sun alter the amount of solar radiation reaching different parts of the planet, triggering long-term climate changes.

However, the transition from an ice age to a warmer period is not a simple, linear process. It involves a complex interplay of feedback mechanisms, some of which amplify the initial warming while others dampen it. For example, as Earth's orbit brings it closer to the sun, melting ice sheets reduce the planet's reflectivity, causing it to absorb more solar radiation and further accelerating warming.

The current warming trend, however, deviates significantly from these natural climate cycles. The rate of warming over the past century is unprecedented in the past million years, far exceeding the pace of natural climate fluctuations. The overwhelming scientific consensus attributes this rapid warming to human activities, particularly the emission of greenhouse gases from burning fossil fuels.

Greenhouse gases, such as carbon dioxide and methane, act like a blanket in the atmosphere, trapping heat and warming the planet. While these gases occur naturally and are essential for maintaining Earth's habitable temperature, human activities have significantly increased their concentrations in the atmosphere, pushing the planet's energy balance out of equilibrium.

The consequences of this rapid warming are already being felt around the world. Sea levels are rising as glaciers and ice sheets melt at an accelerating rate, threatening coastal communities and infrastructure. Extreme weather events, such as heat waves, droughts, and intense storms, are becoming more frequent and severe, impacting agriculture, human health, and ecosystems.

Understanding the interplay between natural climate variability and human-induced warming is essential for predicting the future trajectory of Earth's climate. While natural climate cycles continue to operate, the overwhelming influence of human activities is driving the planet into uncharted territory, with potentially profound consequences for life as we know it.

The urgency to address climate change is underscored by the knowledge that the climate system exhibits inertia. Even if we were to halt all greenhouse gas emissions today, the warming trend would continue for centuries due to the long lifespan of these gases in the atmosphere.

The challenge before us is immense, but not insurmountable. By transitioning to clean energy sources, reducing our reliance on fossil fuels, and adopting sustainable practices, we can mitigate the worst impacts of climate change and create a more sustainable future for generations to come.

Biodiversity Loss The Sixth Mass Extinction

Life on Earth, a tapestry woven over billions of years, is a testament to the power of evolution, adaptation, and resilience. Yet, this intricate web of life is facing an unprecedented crisis, a rapid decline in biodiversity that rivals the great mass extinction events of Earth's deep past. This sixth mass extinction, unlike the previous five, is driven not by natural cataclysms, but by the actions of a single species: our own.

Imagine walking through a lush rainforest, teeming with life, the air alive with the calls of exotic birds, the undergrowth rustling with the movement of unseen creatures. Now picture that same forest, eerily silent, the trees stripped bare, the ground littered with the remnants of a once-thriving ecosystem. This is the stark reality of biodiversity loss, a silent crisis unfolding across the globe, robbing our planet of its natural heritage at an alarming rate.

Scientists estimate that Earth is currently losing species at a rate 100 to 1,000 times faster than the natural background rate of extinction, a pace that surpasses even the most conservative estimates of previous mass extinction events. This rapid decline in biodiversity is akin to losing threads from the tapestry of life, each loss weakening the fabric of ecosystems and jeopardizing the vital services they provide.

The primary driver of this biodiversity crisis is habitat loss and fragmentation, largely driven by agriculture, urbanization, and resource extraction. As forests are cleared, wetlands drained, and grasslands plowed, countless species are losing their homes, their populations squeezed into ever-smaller and more isolated fragments of habitat.

Climate change, another major threat to biodiversity, is exacerbating existing pressures and creating new challenges for species struggling to adapt to a rapidly changing world. As temperatures rise, precipitation patterns shift, and extreme weather events become more frequent, many species are finding their ranges squeezed, their life cycles disrupted, and their survival increasingly precarious.

Overexploitation, whether through overfishing, hunting, or the illegal wildlife trade, is pushing many species towards the brink of extinction. The demand for exotic pets, traditional medicines, and luxury goods is driving a lucrative black market trade that decimates wildlife populations and disrupts ecosystems.

Pollution, in its myriad forms, is another insidious threat to biodiversity. From chemical spills to plastic pollution to the pervasive presence of pesticides and herbicides in our environment, we are contaminating the air, water, and soil upon which life depends.

The consequences of biodiversity loss are far-reaching and profound. Ecosystems, weakened by the loss of species, become less resilient to disturbances, more vulnerable to invasive species, and less able to provide the vital services upon which we depend, such as clean air and water, pollination, and climate regulation.

The loss of biodiversity also has profound cultural and aesthetic implications. Each species represents a unique evolutionary trajectory, a repository of genetic information, and a source of wonder and inspiration. Their loss diminishes our planet's natural heritage and robs future generations of the opportunity to experience the full richness of life on Earth.

The sixth mass extinction is a crisis of our own making, but it is not a foregone conclusion. By addressing the root causes of biodiversity loss, embracing sustainable practices, and fostering a deeper appreciation for the interconnectedness of all life on Earth, we can chart a more hopeful course, one that ensures a thriving planet for generations to come.

The Future of the Quaternary Navigating an Uncertain Path

The Quaternary period, spanning the last 2.6 million years of Earth's history, has been a time of dramatic climate fluctuations, punctuated by ice ages and interglacial periods. Yet, amidst this environmental dynamism, a new force has emerged, one with the potential to reshape the planet and alter the course of evolution: humanity. As we stand at the cusp of the Anthropocene, a new epoch defined by human influence, the future of the Quaternary, and indeed, of life on Earth, hangs in the balance.

Imagine peering into a crystal ball, seeking a glimpse of the future. The image is hazy, obscured by a multitude of possibilities, each determined by the choices we make today. Will we continue down a path of unsustainable consumption, pushing the planet's systems beyond their limits? Or will we chart a new course, one that embraces sustainability, equity, and a deep respect for the natural world?

The challenges we face are daunting, but not insurmountable. Climate change, biodiversity loss, pollution, and resource depletion are interconnected crises that demand urgent action. Yet, within these challenges lie opportunities for innovation, collaboration, and a fundamental shift in our relationship with the planet.

Addressing climate change requires a rapid transition to a low-carbon economy, phasing out fossil fuels and embracing renewable energy sources. This energy transition presents not only an environmental imperative but also an economic opportunity, fostering innovation, creating jobs, and reducing our reliance on finite resources.

Conserving biodiversity requires protecting and restoring natural habitats, combating wildlife trafficking, and addressing the root causes of habitat loss, such as deforestation and unsustainable agriculture. By safeguarding the diversity of life on Earth, we not only preserve the beauty and wonder of the natural world but also enhance the resilience of ecosystems and the vital services they provide.

Reducing pollution necessitates a shift towards a circular economy, one that minimizes waste, promotes recycling, and reduces our reliance on harmful chemicals. By embracing cleaner production processes and adopting sustainable consumption patterns, we can create a healthier planet for both present and future generations.

Addressing resource depletion requires a fundamental shift in our consumption patterns, moving away from a linear "take-make-dispose" model towards a more circular economy that emphasizes reuse, repair, and recycling. By valuing resources not only for their economic worth but also for their ecological importance, we can create a more sustainable and equitable future.

Navigating the uncertain path ahead requires a collective effort, one that transcends national boundaries, political ideologies, and socioeconomic differences. It demands collaboration between governments, businesses, scientists, and citizens, all working together to create a future where human prosperity and environmental sustainability go hand in hand.

Education and awareness are paramount in this endeavor. By fostering a deeper understanding of the interconnectedness of Earth's systems and the impact of human actions, we can empower individuals to make informed choices and demand change from those in power.

The future of the Quaternary is not predetermined. It is a story yet to be written, a tapestry woven from the choices we make today. By embracing sustainability, equity, and a deep respect for the natural world, we can create a future where humanity thrives in harmony with the planet, ensuring a vibrant and resilient Earth for generations to come. The path forward is clear, but the time to act is now.

Chapter 7: The Quaternary Quotient Lessons for the Future

Understanding Past Climate Change Informing Present Action

Earth's climate is a story told in stone, water, and ice. Embedded within ancient coral reefs, layered within polar ice caps, and etched into the rings of towering trees lie clues to past climates, offering a glimpse into a world both familiar and vastly different from our own. By deciphering these natural archives, we gain a deeper understanding of Earth's climate system, its inherent variability, and its sensitivity to external forces, both natural and human-induced.

Picture a time when lush rainforests carpeted Antarctica, their verdant canopies teeming with life. This was the Eocene epoch, some 50 million years ago, a period of exceptional warmth in Earth's history. By studying fossilized plants and analyzing sediment cores, scientists have pieced together a picture of a planet with significantly higher levels of atmospheric carbon dioxide, driving temperatures to levels not seen for millions of years.

Fast forward to the Pleistocene epoch, beginning about 2.6 million years ago, a period marked by dramatic swings between glacial and interglacial periods. These cyclical ice ages, driven by variations in Earth's orbit and solar radiation, left their mark on the planet, carving out valleys, shaping coastlines, and influencing the distribution of plant and animal life.

By analyzing ice cores extracted from the depths of Greenland and Antarctica, scientists can trace Earth's climate back hundreds of thousands of years. Trapped within these frozen time capsules are tiny bubbles of ancient air, preserving a record of past atmospheric composition, including greenhouse gas concentrations.

These paleoclimate records reveal a striking correlation between atmospheric carbon dioxide levels and global temperatures. During warm periods, such as the Eocene, carbon dioxide levels were naturally elevated, while during glacial periods, they were significantly lower. This relationship underscores the crucial role of greenhouse gases in regulating Earth's climate.

The study of past climate change also reveals the interconnectedness of Earth's systems. Changes in ocean circulation patterns, driven by continental drift and variations in solar radiation, have played a significant role in shaping past climates. For example, the formation of the Isthmus of Panama, which connected North and South America, altered ocean currents, contributing to the formation of the Arctic ice cap and ushering in the ice ages of the Pleistocene.

Understanding these past climate transitions provides crucial context for interpreting current warming trends. The rate of warming observed over the past century is unprecedented in the past million years, far exceeding the pace of natural climate fluctuations. This rapid warming, driven by human activities, particularly the burning of fossil fuels, is pushing Earth's climate system into uncharted territory.

The lessons from Earth's past are clear: climate change is not a new phenomenon, but the current rate and magnitude of warming are unprecedented in human history. The burning of fossil fuels is releasing carbon dioxide into the atmosphere at a rate far exceeding any natural process, pushing the planet's energy balance out of equilibrium.

By studying past climate change, we gain a deeper appreciation for the sensitivity of Earth's climate system to external forces, both natural and human-induced. This knowledge is essential for informing present action. By reducing our reliance on fossil fuels, transitioning to clean energy sources, and adopting sustainable practices, we can mitigate the worst impacts of climate change and create a more sustainable future for generations to come. The past is not just a prologue; it is a guidebook for navigating the challenges of the present and shaping a more sustainable future.

Adapting to a Changing World Strategies for Resilience

Change, an intrinsic thread woven into the very fabric of our planet, has been a constant throughout Earth's history. Continents have shifted, climates have fluctuated, and life itself has undergone remarkable transformations, adapting to ever-changing conditions. Yet, the pace of change we are witnessing today, largely driven by human activities, is unprecedented, presenting both challenges and opportunities as we navigate an uncertain future.

Adapting to this changing world requires a fundamental shift in our perspective, one that embraces resilience, innovation, and a deep understanding of the interconnectedness of Earth's systems.

Imagine a city built on stilts, rising above the encroaching tides, its inhabitants living in harmony with the ebb and flow of the sea. This vision of adaptation, once relegated to the realm of science fiction, is becoming a reality in many parts of the world as coastal communities grapple with the impacts of sea-level rise. From building seawalls to restoring coastal ecosystems that act as natural buffers, we are learning to adapt to a changing coastline.

Adaptation is not merely about responding to threats; it is about embracing opportunities and building resilience into our systems. In agriculture, for example, farmers are adopting drought-resistant crops, implementing water conservation techniques, and diversifying their yields to cope with changing precipitation patterns and increasing temperatures.

Building resilient infrastructure requires incorporating climate projections into design standards, ensuring that roads, bridges, and buildings can withstand more extreme weather events. Cities are creating green spaces to mitigate the urban heat island effect, implementing permeable pavements to reduce flooding, and investing in public transportation systems that reduce reliance on private vehicles.

Adapting to a changing world also necessitates addressing social vulnerability. Climate change exacerbates existing inequalities, disproportionately impacting marginalized communities who often lack the resources to cope with extreme weather events, sea-level rise, and other climate-related challenges. Ensuring equitable access to resources, information, and decision-making processes is crucial for building a just and resilient future.

Nature-based solutions, inspired by the inherent resilience of ecosystems, offer a powerful approach to adaptation. Restoring coastal wetlands, for example, not only provides a buffer against storm surges and sea-level rise but also enhances biodiversity, improves water quality, and sequesters carbon dioxide.

Investing in early warning systems, particularly in regions prone to natural disasters, can save lives and reduce economic losses. By providing timely and accurate information about impending hazards, communities can take proactive measures to prepare and respond effectively.

Adaptation is not a one-size-fits-all solution. Strategies must be tailored to specific geographic locations, social contexts, and environmental challenges. Local communities, with their intimate knowledge of their environments and vulnerabilities, are essential partners in developing and implementing effective adaptation measures.

The transition to a more resilient future requires a fundamental shift in our values and behaviors. Embracing sustainable consumption patterns, reducing our reliance on fossil fuels, and investing in renewable energy sources are not only essential for mitigating climate change but also for building a more equitable and prosperous future.

Adapting to a changing world is not an option; it is an imperative. By embracing innovation, collaboration, and a deep understanding of the interconnectedness of Earth's systems, we can navigate the challenges ahead and create a more resilient and sustainable future for generations to come. The path forward is clear: adapt, innovate, and thrive in the face of change.

The Importance of Interdisciplinary Research Connecting the Dots

The natural world, a symphony of interconnected systems, defies categorization. From the intricate dance of molecules within a single cell to the complex interplay of organisms within an ecosystem, life on Earth is a testament to the interconnectedness of all things. Understanding the complexities of our planet, particularly in the face of unprecedented environmental change, requires a holistic approach, one that transcends the traditional boundaries of scientific disciplines. Interdisciplinary research, the weaving together of diverse perspectives, methodologies, and expertise, is not merely an option; it is the key to unlocking solutions for a sustainable future.

Imagine a team of scientists, their expertise spanning fields as diverse as climatology, ecology, economics, and social science, gathered around a table, their collective focus on a single challenge: understanding the impacts of climate change on a coastal community. This collaborative endeavor, fueled by a shared sense of purpose, exemplifies the power of interdisciplinary research.

Consider the challenge of predicting the future of coral reefs, vibrant ecosystems facing multiple threats from ocean acidification, warming waters, and pollution. A marine biologist, armed with an understanding of coral physiology and the intricate web of life that depends on these underwater cities, can assess the direct impacts of these stressors. However, understanding the broader social and economic implications requires a broader lens.

An economist can shed light on the economic value of coral reefs, from tourism revenue to coastal protection, quantifying the potential costs of their decline. A social scientist can provide insights into the cultural significance of reefs to local communities, highlighting the potential social and political ramifications of their loss. By integrating these diverse perspectives, a more comprehensive understanding emerges, one that informs effective conservation strategies.

The study of past climates, a field known as paleoclimatology, exemplifies the power of interdisciplinary research. Reconstructing Earth's climate history requires piecing together evidence from a wide range of sources, each requiring specialized expertise. Geologists analyze sediment cores extracted from the ocean floor, deciphering the chemical signatures of ancient oceans. Paleontologists study fossilized plants and animals, reconstructing past ecosystems and their responses to climate change.

Ice core scientists, working in some of the most extreme environments on Earth, extract frozen time capsules from glaciers and ice sheets, analyzing trapped air bubbles to reconstruct past atmospheric composition. By combining these diverse lines of evidence, paleoclimatologists can paint a detailed picture of Earth's climate history, providing crucial context for understanding present-day warming trends.

The challenges facing our planet are complex and interconnected, demanding integrated solutions that transcend disciplinary boundaries. Addressing climate change, for example, requires not only reducing greenhouse gas emissions but also adapting to the impacts that are already occurring. This multifaceted challenge demands collaboration between engineers developing renewable energy technologies, policymakers crafting effective climate policies, and social scientists understanding the barriers to behavioral change.

Interdisciplinary research, however, faces its own set of challenges. Breaking down disciplinary silos requires overcoming institutional barriers, fostering a culture of collaboration, and developing a shared language that bridges disciplinary divides. Funding agencies play a crucial role in fostering interdisciplinary research by supporting projects that transcend traditional boundaries and encouraging collaboration across disciplines.

The future of scientific discovery lies in embracing the interconnectedness of our world. By fostering interdisciplinary research, we can unlock the full potential of human ingenuity, harnessing the collective wisdom of diverse minds to address the complex challenges facing our planet and create a more sustainable future for all. The dots are there; it is through interdisciplinary research that we can connect them, revealing the bigger picture and charting a course toward a brighter future.

The Quaternary A Legacy for Future Generations

The Quaternary period, spanning the last 2.6 million years of Earth's history, stands as a testament to change and resilience. It is a period defined by dramatic climate fluctuations, from the frigid grip of ice ages to the relative warmth of interglacial periods, a rhythmic pulse that has shaped landscapes, influenced the course of evolution, and left an indelible mark on the planet we inherit today. Understanding the legacy of the Quaternary, particularly the interplay between climate, ecosystems, and human evolution, is crucial for navigating the challenges of the Anthropocene, a new epoch defined by human influence on Earth's systems.

Imagine a world encased in ice, vast glaciers stretching across continents, their icy fingers carving out valleys and shaping coastlines. This was the Pleistocene epoch, a time of intense cold punctuated by brief interglacial respites. Yet, even in this seemingly inhospitable world, life persisted, adapting and evolving to survive the challenges of a changing climate.

The Quaternary witnessed the rise of our own species, Homo sapiens, a testament to the adaptability and resilience of life. Our ancestors, faced with the challenges of a fluctuating climate, developed sophisticated tools, harnessed the power of fire, and formed complex social structures, laying the foundation for the development of human civilization.

The legacy of the Quaternary is etched into the very fabric of our planet. The rich soils of the American Midwest, for example, owe their fertility to the grinding action of glaciers during the last ice age, which deposited vast quantities of mineral-rich sediment. The Great Lakes, a defining feature of the North American landscape, were carved out by the advance and retreat of massive ice sheets.

The distribution of plant and animal life today is a direct reflection of the Quaternary's climatic upheavals. Species migrated across continents in response to changing temperatures and shifting vegetation patterns, leading to the diverse ecosystems we see today. Understanding these past migrations can inform conservation efforts in the face of future climate change.

The Quaternary also holds lessons about the interconnectedness of Earth's systems. Ice core records reveal a tight coupling between atmospheric carbon dioxide levels, global temperatures, and ice sheet extent. These records serve as a stark reminder of the sensitivity of Earth's climate system to even small changes in greenhouse gas concentrations.

The rise of agriculture, a pivotal moment in human history, occurred during a period of relative climate stability following the last glacial maximum. This stable climate provided the conditions necessary for the development of settled agriculture, which in turn led to the rise of cities and the emergence of complex societies.

However, the legacy of the Quaternary is not without its cautionary tales. The extinction of megafauna, such as woolly mammoths and saber-toothed cats, coincided with the arrival of humans in various parts of the globe, highlighting the potential impact of our species on the natural world.

As we navigate the challenges of the Anthropocene, the Quaternary serves as both a guide and a warning. It reminds us of the resilience of life, the interconnectedness of Earth's systems, and the profound influence of climate on the course of evolution. It also highlights the potential consequences of human actions on the planet we call home.

By learning from the past, we can make more informed decisions about the future. By understanding the legacy of the Quaternary, we can better appreciate the delicate balance of Earth's systems and work towards a more sustainable future for generations to come. The Quaternary is not just a chapter in Earth's history; it is a prologue to the future we are writing today.